Raised Hands, Hidden Stories:

A Teacher's View from Inside the Classroom

by

Dr. Leonie M. Curtis-Brooks

Introduction: Behind the Classroom Door

Every classroom is alive with stories, some you can spot right away, while others you'll only notice if you look carefully enough. Some kids raise their hands for every question, while others try not to catch your eye at all. Some speak up, solving problems at the board with practiced confidence, and the ones who seem content to take up as little space as possible. Some students are shy, whereas others are more confident.

As a teacher, it's easy to focus on the obvious: the students who demand your attention. But over the years, I've learned that the quieter stories, the ones hidden behind lowered eyes and silent desks, matter just as much.

My name is Dr. Leonie M. Curtis-Brooks, and for over a decade, I've taught mathematics in high school classrooms filled with every kind of student you can imagine. When I started, I thought teaching was mostly about the subject clarifying equations, grading homework, and explaining concepts. But the longer I've stayed, the more I've realized that the real work of teaching happens beyond the curriculum. It's about noticing what's unspoken. It's about understanding the small ways students cope, the moments when a glance or a sigh tells a bigger story than any test score ever could.

There's a kind of invisibility that settles over some students. I've met kids who sit in the back row, never volunteering, never causing trouble, almost fading into the walls. At first, you might assume they don't care or that they're unmotivated. But if you pay attention, you start to see the truth: many kids carry burdens they never mention. Learning difficulties, family chaos, language hurdles, or just the slow, grinding work of growing up in a world that sometimes seems rigged against them. This book is all about those students, the ones whose stories don't make the yearbook or the honor roll, but whose daily struggles and quiet victories are just as profound.

Teaching, at its core, is about seeing what others miss. It's about listening to what isn't said, and believing that every student, even the ones who never raise their hands, has a story worth hearing. I've watched students succeed in ways that never show up in their GPA, finishing an assignment they thought was impossible, asking a question after weeks of silence, or simply making it through a tough day. These small wins are enormous, even if they never show up on a transcript.

This book is a collection of those moments. It's not a manual or a list of strategies, though I hope you'll find some useful ideas here. It's a series of stories, vignettes from the front lines of teaching, where academic achievement and personal struggle are always tangled together. It's also a plea: for teachers to look beyond the obvious, for parents to ask deeper questions, for students to know they're seen and valued even when they feel invisible.

The classroom is more than a place for lessons and tests. It's a microcosm of life, where kids learn not just about fractions or grammar, but about themselves. Behind every raised hand is a story, and behind every hesitation is a reason. I hope that by sharing what I've seen, the quiet courage, the daily resilience, the hidden pain, and unexpected joy, I can help others see their students, their children, and maybe even themselves, a little more clearly.

So, welcome to the classroom as I know it: messy, unpredictable, full of heartbreak and hope. The stories you'll read here are real, though names and details have been changed to protect privacy. They are stories of struggle, of small triumphs, and of the deep, invisible work that happens whenever someone takes the time to notice. My greatest wish is that, after reading, you'll notice a little more, too.

Table of Contents

Chapter 1
The Students You Don't See

The first time I realized I was missing something important, I was halfway through a lesson on quadratic equations. I'd been teaching for a few years by then, and I had my routine: pose a problem, scan the room for raised hands, and pick the eager faces. It felt efficient. But one day, as I circled the room, I caught sight of Maria, quiet Maria, always in her seat by the window, always looking down, never volunteering. Her notebook was filled with neat handwriting and half-finished problems, and her pencil tapped the margin as if she were working up the nerve to write something down.

"Maria, do you want to walk us through the first step?" I asked, hoping to nudge her forward. She barely looked up, then shook her head, almost apologetically. I moved on, but I couldn't stop thinking about her silence. Was it shyness? Disinterest? Or something else entirely?

Weeks passed, and I started to pay more attention. I began to notice the constellation of quiet kids in every class, students who never caused trouble, who never interrupted, and who sometimes seemed to be holding their breath to get through the period unnoticed. I started to wonder: What was I missing? What stories were sitting right in front of me, hidden behind lowered eyes and still hands?

I started small, asking gentle questions, leaving notes on their desks: "I see you working hard today. Keep it up." Sometimes, I'd catch a flicker of surprise, a shy smile, or a little more effort the next day. Other times, I got nothing at all. But little by little, some of those students started to open up.

The Perfectionist

Sophie sat in the third row, always early, always prepared. Her binder had color-coded tabs. Her homework was never late. But she never, ever raised her hand. Not once in the entire first semester.

At first, I assumed she was just introverted, content to learn quietly, absorbing everything without needing to perform. But during a group quiz review, I noticed something odd. When her partner got an answer wrong, Sophie's whole body tensed. She gripped her pencil so hard her knuckles went white. When it was her turn to share an answer, she whispered it so quietly that her partner had to ask her to repeat it three times. Even then, she added, "But I'm probably wrong."

She wasn't wrong. She was rarely wrong.

I started watching more carefully. Sophie would complete a problem, then erase it. Complete it again, erase it again. Sometimes she'd go through this cycle four or five times, each answer virtually identical to the last. During tests, she'd finish early but sit frozen at her desk, eyes scanning her paper over and over, too afraid to turn it in. I'd have to take it from her when time was up, gently.

One day, I asked her to stay after class. She looked terrified, like I was about to tell her she was failing.

"Sophie, you're doing really well in this class," I started. "But I've noticed you seem... stressed. Is everything okay?"

Her eyes filled with tears immediately. "I just can't mess up. I can't. My parents expect " She stopped, swallowing hard. "I got a B+ on the last test, and my mom didn't talk to me for two days."

There it was. The weight she'd been carrying was invisible to everyone around her. The silence wasn't shyness or disinterest. It was fear. Fear of being wrong, of being imperfect, of disappointing someone.

I tried everything I could think of. I praised her process, not just her results. I told her that mistakes were how we learned, that even I sometimes got problems wrong. I gave her intentionally challenging problems, where getting stuck was the point. I encouraged her to ask questions, reminding her that asking questions was a sign of thinking, not weakness.

Some days, it seemed to help. She'd raise her hand tentatively, voice shaking, and ask for clarification. I'd celebrate it as if she'd just solved Fermat's Last Theorem. Other days, she'd retreat into herself, erasing and re-erasing, trapped in her own head.

By the end of the year, Sophie was still quiet. But occasionally, just occasionally, she'd leave a problem with a small mistake uncorrected, or she'd turn in a test without checking it seventeen times. Once, she even laughed when she caught an error during group work, saying, "Oops, I always mix those up." It wasn't a transformation. But it was something, a tiny crack in the armor of perfectionism.

The Outsider

Then there was Lin. She'd moved from China at the start of the school year, and her English was functional but halting. She sat at the back corner, as far from the front as possible, and rarely made eye contact with anyone. During group work, other students would talk around her, not unkindly, but as if she wasn't quite there.

I assumed the issue was language. I got her some translated materials, paired her with patient classmates, and spoke to her slowly and clearly. She'd nod politely, write things down, and remain silent.

But one day, I noticed her notebook was filled with work far beyond what we were covering in class. While we were working on linear equations, she had pages of trigonometric identities, calculus notation, and complex proofs. The work was sophisticated, well beyond grade level.

"Lin," I said, kneeling by her desk, "this is really advanced. Did you study this in China?"

She nodded, barely meeting my eyes.

"So... is this class too easy for you?"

A long pause. Then, quietly: "Yes. But I don't know how to say. My English is not good."

My heart sank. Here I'd been, assuming she was struggling, when really she was bored out of her mind. But it wasn't just about the math. It was about everything being in a new country, a new school, where she didn't know the social rules, didn't understand half the cultural references, and didn't have friends. The math was easy. Everything else was impossibly hard.

I started giving her extension problems, advanced challenges she could work on independently. I introduced her to another student, Kevin, who was also ahead in math and was learning Mandarin. I made a point of asking her to share her problem-solving methods with the class—not in front of everyone at first, but by writing them on the board before class started so that students could see different approaches.

Slowly, something shifted. Kevin and Lin started sitting together, working through competition math problems during lunch. Other students began asking her for help, and she'd explain things carefully, drawing diagrams when words failed her. Her English didn't improve overnight, but her presence did. She took up more space. She smiled more.

One day, during a particularly tricky problem, she raised her hand. "I think... maybe there is a different way?" she said, her accent thick but her voice clear. She walked to the board and showed us an elegant solution I'd never seen before, one she'd learned in her previous school.

The class was silent for a moment. Then someone said, "Whoa, that's so much faster."

Lin ducked her head, but she was smiling.

Her quietness didn't disappear; she was still reserved, still careful with her words. But it was a different kind of quiet now. Not the silence of invisibility, but the quiet of someone who was choosing when to speak.

The Unreachable One

But not every story has a happy ending. Not every quiet student opens up, no matter what you try.

Marcus was a junior who'd transferred mid-year. He sat in the back, hood up, earbuds in until the bell rang. He did the bare minimum enough to pass, never enough to stand out. He didn't cause problems, didn't talk back, didn't engage. He was a ghost.

I tried everything I'd learned. I left encouraging notes on his papers. I asked him to stay after class to check in. I offered extra help, extensions, and alternative assignments. I tried humor, tried sincerity, tried giving him space, tried giving him attention.

Nothing worked.

"Marcus, I noticed you didn't finish the homework. Do you need help with it?"

"Nah, I'm good."

"How are things going? Anything I can do to support you?"

"I'm fine."

"I saw you got a low score on the quiz. Want to do a retake?"

"Maybe later." (He never did.)

I talked to his other teachers. Same story everywhere. I reached out to his counselor, who told me he'd been through three schools in two years, that his home life was "complicated," and that he'd learned not to get attached to places or people. I tried to keep that in mind, tried to be patient, tried to be consistent.

But Marcus had built walls I couldn't scale. Every attempt I made seemed to bounce off him. He'd look at me with flat, tired eyes that said, *I know what you're doing, and it won't work.*

One day, frustrated after another one-word conversation, I asked him directly: "Marcus, what do you need from me? How can I help you succeed in this class?"

He looked at me for a long moment. Then he said, "You can't."

It wasn't said with anger or defiance. It was just... factual. Resigned. Like he'd already decided that school, teachers, all of it—it didn't matter. He was putting in time until he could leave.

I kept trying, but more quietly. I stopped pushing for conversations he didn't want to have. I made sure he knew where I was, in case he needed me. I graded his work fairly but didn't penalize him for late submissions. I let him sit in the back with his hood up, as long as he wasn't disrupting anyone.

He passed my class with a C-. On the last day of school, he turned in his final exam, and as he walked past my desk, he said, "Thanks."

Just that. One word. I don't know what he was thanking me for—maybe for not giving up entirely, or maybe for finally leaving him alone. I don't know if I helped him at all. I don't know if I failed him by not trying harder, or if trying harder would have just pushed him further away.

What I learned from Marcus is that sometimes, you can't reach a student. Sometimes, the trauma is too deep, the walls too high, the exhaustion too complete. Sometimes, a kid has been let down so many times that your kindness feels like a trick, and trust feels more dangerous than isolation.

It's a hard lesson. It goes against everything we want to believe as teachers that if we care enough, try hard enough, we can reach every student. But the truth is messier than that. Some students aren't ready to be reached. Some students have survival mechanisms that look like apathy but are, in fact, armor. And sometimes, the best you can do is plant a seed and hope that someday, maybe years later, it grows into something.

I still think about Marcus. I hope he's okay. I hope that somewhere along the way, someone got through. Or maybe he got through on his own. I'll probably never know.

The Hidden Genius

And then there was Aisha, who taught me that quiet doesn't always mean struggling—sometimes it means the opposite.

Aisha was so quiet I sometimes forgot she was in the room. She'd sit near the middle, never in the front, never in the back, perfectly camouflaged in the sea of students. She turned in her homework on time, scored decently on tests—solid B's, nothing remarkable. She never asked questions, never needed help, never drew attention to herself.

I probably wouldn't have noticed her at all if I hadn't accidentally seen her doodling during a lecture. I was about to say something gently, because she wasn't being disruptive—when I realized she wasn't doodling. She was working through a proof. A complex, elegant proof of the theorem I was teaching, using methods I hadn't shown the class yet.

I stopped mid-sentence. "Aisha, what are you working on?"

She looked up, startled, and quickly covered her notebook. "Nothing. Sorry."

"No, don't apologize. Can I see?"

Reluctantly, she slid the notebook toward me. The page was covered in mathematical notation, diagrams, and alternative proofs. She'd been extending the lesson on her own, exploring implications I hadn't even mentioned.

"This is... really sophisticated work," I said. "Have you studied this before?"

She shrugged. "I just like math. I do it for fun sometimes."

For *fun*. While I was teaching concepts she'd already mastered, she was entertaining herself by diving deeper, finding patterns, and making connections. And she'd never said a word.

"Why didn't you tell me you were this far ahead?"

Another shrug. "I don't know. I didn't want to seem like I was showing off. And it's nice, just... doing it quietly. Without pressure."

I realized then that Aisha's silence was a choice. She wasn't struggling, scared, or disconnected. She was content. She'd found a way to engage with the material on her own terms, without the performance anxiety that came with being labeled "gifted" or "advanced." She didn't need validation or attention. She just needed space to think.

I made her a deal. She could work on whatever she wanted during class, as long as she checked in with me once a week to show me what she was exploring. I gave her access to competition math resources, college-level textbooks, and online courses. I didn't make a big deal about it in front of the class. I just let her be.

Every Friday, she'd stay for a few minutes after class, and we'd talk about what she'd discovered that week. Her eyes would light up as she explained some elegant solution or surprising connection. She was a completely different person in those moments—animated, excited, fully alive.

She stayed quiet in class. But it was a peaceful, quiet now, the quiet of someone deeply engaged in their own intellectual journey. She didn't need to raise her hand or prove herself to anyone. She was learning for the pure joy of it.

At the end of the year, she gave me a handwritten note. "Thank you for letting me be invisible," it said. "Sometimes that's exactly what I needed."

The Long Game

Jamal was one of them. He rarely turned in homework on time, and his grades swung wildly from week to week. I used to think he was just lazy. But one afternoon, I caught him in the hallway and asked how things were going. He hesitated, then told me, in a voice so low I almost missed it, that his parents worked nights and he was in charge of his younger siblings. He spent most evenings cooking, cleaning, and making sure everyone got to bed. Schoolwork happened whenever he could find a minute. Suddenly, the late homework and tired eyes made sense.

After that, I tried to meet Jamal where he was. I gave him extensions when he needed them, checked in quietly, and let him work in my classroom after school when he could. He didn't transform overnight; life doesn't work that way, but he started to trust me. One day, during group work, I saw him explaining a problem to a classmate, his voice steady and sure. That moment, as small as it was, felt like a breakthrough.

Then there was Elias, who barely spoke at all. I paired him with a patient classmate for a project, and at first, I thought it would flop. But slowly, Elias started to contribute first a few words, then full sentences, then ideas of his own. By the end of the week, he was laughing with his partner, proud of what they'd built together.

But it's not just about the breakthrough moments. Sometimes, it's about the days when nothing seems to change—when a student remains quiet, when eye contact is brief, when the wall between you and them seems as solid as ever. That's when persistence matters most. I learned to watch for the invisible ways students participate. Some kids shine quietly, helping a friend or just showing up every day despite what they carry outside of school. I learned to celebrate effort, not just answers. I started leaving more notes, giving small jobs, passing out papers, and organizing supplies for the kids who never volunteered. Sometimes, all it took was a little attention to draw them out.

I also learned that the classroom itself matters. Environments that reward only the loudest voices can leave quieter kids behind. So, I started creating more low-pressure ways to participate: writing answers on sticky notes, discussing with a partner before sharing with the group, and letting students submit questions anonymously. I made it clear that mistakes were expected, that trying was more important than being right.

Every class is full of hidden stories. Some students are surviving things you'll never know about: hunger, fear, loneliness, and the feeling that they don't belong. Others are just trying to make it to the next day. The smallest moments finishing a worksheet, asking a question, even just showing up—can be victories. As a teacher, your job is to notice, to care, and to let kids know they matter, whether or not they ever raise their hand.

Looking back, I realize how many students I almost missed, how many stories I never would have known if I hadn't bothered to look. The invisible kids are always there, waiting for someone to see them. And when you do, you realize how much they have to teach you about resilience, about hope, and about the quiet strength it takes to keep going.

Sometimes, the truest reward isn't a raised hand or a perfect test score. It's the slow, steady emergence of a student who once sat invisible. It's a note left behind after graduation, or a quiet "thank you" in the hallway. It's the knowledge that, just by noticing, you've given someone the permission to be seen.

And sometimes, the reward is simply knowing you tried even when it didn't work, even when the student walked away still silent, still struggling. Because trying matters. Noticing matters. Even when you can't fix everything, even when you can't reach everyone, the act of seeing a student really seeing them plants something. Maybe it grows right away. Maybe it takes years. Maybe it never grows at all.

But you tried. And in a world where so many kids feel invisible, that has to count for something.

Reflective Prompts for Educators:

• How can I truly uncover the hidden strengths of my quietest students? What can I do to spark their confidence and make them feel seen?

• What creative ways can I introduce small gestures of recognition to help even the most reserved students break through their silence and participate?

• Is my classroom environment a welcoming space for those who tend to fade into the background? How can I transform it into one that invites even the most withdrawn to engage?

• What strategies can I put into action today to ensure that every single student, no matter how quiet, feels recognized, heard, and valued in every moment?

- Am I fully ready to accept that some students may not respond immediately to my efforts? How can I continue pushing forward with enthusiasm while respecting their personal boundaries?

- What unique kinds of "quiet" do I see in my students, and how can I tailor my approach to meet the needs of each?

Chapter 2
Brilliance Under Pressure

Every teacher has met a student whose mind moves like lightning—who can see patterns, make leaps, and ask questions that stop you in your tracks. But sometimes, those students are also the ones whose potential is buried under a mountain of pressure. Sometimes, brilliance has to fight for space.

Take Julio. He could solve complex problems in his head and explain concepts in ways that made me rethink my own lesson plans. But his grades were a rollercoaster: A+ one week, D the next. He'd come in looking exhausted, eyes rimmed red, and sometimes he'd drift off right in the middle of class. I used to pull him aside and ask, gently, "Are you getting enough sleep?" He'd grin sheepishly and admit he'd been up late—sometimes because of video games, sometimes because his house was loud and chaotic, sometimes just because he couldn't quiet his mind.

What struck me most about Julio was the disconnect between his ability and his performance. He could articulate complex ideas in casual conversation, yet when it came time to write a test, something shut down. I watched him stare at the page, his leg bouncing with nervous energy, his pencil hovering but never quite touching paper. After he'd turn in a half-finished exam, I'd find him in the hallway, relaxed and joking, explaining the very concepts he couldn't access under pressure. It was as if his brilliant mind had two modes: creative and free, or frozen and afraid.

Over time, I realized that Julio's struggle wasn't about intelligence. It was about the weight he carried. His older sister had dropped out of college, and his parents constantly reminded him that he had to "make it"—that their sacrifices depended on his success. At home, his younger siblings looked up to him, expecting him to have all the answers. At school, teachers had high expectations, sometimes expressed with subtle pressure: "You're so smart, Julio, why aren't you applying yourself?" That phrase, meant as motivation, often felt like judgment to him.

Julio's story is more common than many realize. There are so many pressures that crowd a teenager's life: expectations from family, the pull of friendships, the lure of escapism, and the relentless hum of social media. For some, it's a job after school, or caring for siblings, or trying to fit in with a group that expects you to hide your intelligence. For others, it's a fear of letting someone down themselves, their parents, or even their teachers. There's also the internal pressure that brilliant kids place on themselves, the perfectionism that whispers that anything less than excellence is failure.

There was a year when my class was full of these quiet strugglers. There was Priya, who was so afraid of making mistakes that she'd freeze up anytime I asked her to show her work on the board. She'd volunteer answers, but the moment I asked her to write them publicly, her hand would shake, and her mind would go blank. I learned later that her father had a habit of criticizing her work harshly, pointing out every error, no matter how small. For Priya, a mistake wasn't just a mistake; it was evidence of inadequacy. There was Malik, who was so busy supporting his family that homework felt like a luxury. He worked part-time at a grocery store to help with rent, came home to help his younger siblings with homework, and still found time to attend my class. Yet his grades suffered because there simply weren't enough hours in the day. And there was Emily, who was brilliant in discussion but crumbled on tests, paralyzed by anxiety. She could articulate sophisticated ideas in a group setting, but the moment the pressure intensified and the stakes felt high, her mind went blank. She'd leave exams in tears, not because she didn't know the material, but because her anxiety had hijacked her ability to access it.

For all of them, pressure didn't just come from one source. It came from everywhere, a chorus of voices, real and imagined, telling them who they should be, what they should achieve, and how they should act. Some students responded with rebellion, pushing back against the weight and making poor choices. Others retreat, withdrawing into themselves and giving up. Some hid their struggles behind jokes, or behind bravado, or behind a carefully constructed mask of indifference. What was heartbreaking was that these defenses often made things worse. A student who joked their way through class wasn't engaging with the material. A student who appeared indifferent was often deeply hurt by their own perceived failures.

I began to see the patterns more clearly. The brilliant kids who weren't performing weren't lazy—they were suffocating. And the more pressure we applied, the more they struggled. It was like watching someone hold their breath underwater: you can only stay down so long before you have to come up for air, gasping.

Understanding the Architecture of Pressure

What I learned through years of working with students like Julio, Priya, Malik, and Emily is that academic pressure operates on multiple levels. There's the explicit pressure: the grades that matter, the college applications that loom, the test scores that determine opportunities. But there's also the implicit pressure, the unspoken messages that students internalize from the people around them. When a parent says, "I know you can get an A in that class," the student hears not just encouragement, but expectation. When a teacher says, "You're so smart, I know you'll figure this out," the student might hear, "I'm disappointed when you don't."

Then there's the internal pressure, often the most destructive. Brilliant students frequently develop perfectionism as a shield. If I can't be perfect, they think, then I shouldn't try at all. This all-or-nothing thinking becomes a prison. They set impossible standards for themselves, and when they inevitably fall short, they internalize the failure as a character flaw rather than a human reality.

I also came to understand that pressure affects different students in different ways. Some become anxious and avoidant, like Emily. Others become aggressive and defiant, like some of Julio's friends who started skipping class. Some become people-pleasers, overextending themselves and burning out. And some, like Priya, become hypervigilant, constantly monitoring for signs that they're failing.

Strategies That Actually Work

I learned that reaching these students wasn't about pushing harder. It was about pulling back, about making space for honesty and vulnerability. With Julio, I broke assignments into smaller pieces and checked in regularly—not to scold, but to ask, "How can I help?" This simple shift changed everything. Instead of waiting for him to fail, I'd catch struggles early. More importantly, I was signaling that I cared about his well-being, not just his grade.

I also explicitly talked to Julio about pressure. I told him about my own experiences with perfectionism; about times I'd frozen up or given up because the stakes felt too high. I shared that even now, there are things I avoid because I'm afraid I won't be good at them. This vulnerability mattered. It showed him that his struggles weren't unique or shameful, and that highly capable people also grapple with pressure.

With Priya, I offered private feedback and celebrated her risk-taking, not just her right answers. I invited her to present her work to me one-on-one before showing it to the class. This gave her a low-stakes opportunity to practice and build confidence. More importantly, I praised her effort and her willingness to try, not just the quality of the final product. Over time, her willingness to take risks grew, and with it, her performance improved.

With Malik, I worked out flexible deadlines and praised every step he took towards catching up, no matter how small. I also connected him with a tutor who understood his time constraints. But beyond the logistics, I made sure Malik knew that I saw him, that I recognized how hard he was working, how much he was juggling. Sometimes, that recognition was more powerful than any grade improvement.

With Emily, I offered alternative ways to demonstrate her understanding. Instead of traditional timed tests, I let her do oral presentations, write a detailed explanation of solving a problem, or create projects. I also taught her specific anxiety management techniques: deep breathing, positive self-talk, and ways to reframe test anxiety as excitement. We practiced these together, starting small and building up.

The Courage to Fail

Sometimes, it took courage to let students fail safely. I started offering low-stakes opportunities to make mistakes. Students could revise assignments, retake quizzes, and present work in progress without penalty. This reframed failure not as an ending, but as part of the learning process. When Julio turned in a poorly reasoned step-by-step process of a question, I didn't mark it down harshly. Instead, I asked him questions that helped him see where his thinking had gone astray, and then I gave him the chance to revise. Over time, he came to see revision not as evidence of initial failure, but as a normal part of growth.

Sometimes, it was about helping them see that their worth wasn't tied to a grade or a single performance. I started having one-on-one conversations with students where I explicitly said things like: "Your value as a person has nothing to do with this test score. You are smart, capable, and worthy, regardless of what happens in this class." These conversations felt radical in a system obsessed with measurement and ranking. But they mattered. Students needed to hear that their teachers saw them as whole people, not just as academic performers.

Sometimes, it was about connecting them with resources outside my classroom, such as counselors, mentors, and support groups. I realized that as a teacher, I had limits. If Julio was dealing with family pressure or trauma, that was beyond what I could address alone. But I could help him access people who could. I learned to have conversations with school counselors, to recommend community programs, and to help students see that asking for help was a sign of strength, not weakness.

The Hidden Costs of Pressure

What I've come to understand is that unmanaged pressure doesn't just affect grades. It affects the whole student. Kids under constant pressure develop anxiety, depression, and physical symptoms like headaches and stomach problems. They become so focused on the external markers of success, grades, test scores, and college applications that they lose touch with their own curiosity and joy in learning. Julio, who could have loved problem-solving, came to hate it because it felt like a test he was failing. That's a tragedy.

I also noticed that pressure disproportionately affects already-vulnerable students. Malik was under pressure not just to succeed academically, but to lift his family out of poverty. Priya was under pressure to be the "good daughter" who made her parents' sacrifices worthwhile. Emily was under pressure because her anxiety made ordinary tasks feel insurmountable. These students were already starting from a disadvantaged position, and society's response was often to push harder, to demand more, to tell them to "just work harder" or "stop being so nervous."

What I Learned Is That Brilliance Under Pressure

What I learned is that brilliance under pressure doesn't always look like success. Sometimes, it looks like survival. Sometimes, it looks like a student showing up after a hard night, or trying again after a failure, or asking for help for the first time. These aren't the moments that get celebrated in assemblies or posted on social media. But they're profound acts of courage.

I also learned that the most important thing I could do was help students develop resilience, not by pushing them harder, but by helping them build skills and mindsets that could weather pressure. This meant teaching stress management techniques, helping them develop a growth mindset, and modeling the ability to bounce back from failure.

Our job as educators isn't just to challenge students; it's to create an environment where they can admit they're struggling, where they can try and fail and try again. This requires a fundamental shift in how we think about pressure and performance. Instead of seeing pressure as motivating, we need to recognize when it becomes toxic. Instead of praising students for how much they can endure, we need to help them develop healthy boundaries. Instead of celebrating the students who appear to have it all figured out, we need to make space for those who are struggling openly.

The light inside a student like Julio is real, but it can be fragile. It needs protection, patience, and the belief that, no matter how many setbacks, it's worth nurturing. When we see the whole student brilliance, pressure, and all, we give them a fighting chance to shine.

Reflective Prompts for Educators:

- How can I create a classroom environment that reduces pressure while still challenging students to grow and succeed?

• What small steps can I take to ensure that students like Julio and Priya feel supported and understood, rather than judged or overwhelmed?

• How can I help my students recognize that their value isn't tied to their grades or academic performance, but to their intrinsic worth and potential?

• In what ways can I actively listen to my students' struggles and offer resources or support that extend beyond the classroom to help them cope with external pressures?

• How can I cultivate a mindset in my students that views failure as a natural part of growth, not something to fear or avoid?

• Am I creating enough opportunities for students to engage with their learning process without the fear of judgment or failure? How can I make space for them to express vulnerability?

• How can I address the unique pressures my students face—whether academic, familial, or social—and ensure that my expectations align with their realities?

• What actions can I take to ensure that brilliant students who struggle with pressure receive the support they need to thrive, both academically and emotionally?

• Am I actively encouraging my students to seek help when needed and reassuring them that it's okay to ask for support?

• What can I do to help students build resilience and stress-management skills, not just for the classroom, but for life beyond school?

Chapter 3
Beyond Grades

If you spend enough time in a classroom, you start to see how much life happens outside the boundaries of a report card. Grades might be the way we track progress, but they're a blunt instrument good for sorting, bad for seeing. A grade tells you if a student got the right answer. It doesn't tell you if they tried for the first time, if they helped a struggling classmate, if they overcame their fear to ask a question, or if they spent hours wrestling with an idea that ultimately led them somewhere new.

The obsession with grades is pervasive in modern education, and I don't blame teachers for it. We operate in a system that demands quantifiable measures of success. Colleges want GPAs. Parents want to see A's and B's. The bureaucracy requires data. Yet something essential gets lost in this translation of learning into numbers. We reduce the messy, multifaceted process of intellectual growth into a single digit or letter. We tell students that an A means they've succeeded and a C means they've failed, when the reality is far more nuanced.

Anaya was a case in point. She was always curious; her hand went up not just for answers, but for questions. She'd look at a problem from every angle, sometimes getting so wrapped up in her own theories that she'd forget to write anything down. Her grades hovered in the middle, not failing, but never drawing much notice. But when we did open-ended projects, her creativity was dazzling. She'd invent new ways to approach a problem, drawing diagrams in the margins, connecting ideas that weren't obvious to anyone but her. Her project on renewable energy sources, for instance, was a work of art, visually stunning, conceptually sophisticated, and entirely her own creation. Yet because it didn't fit the rubric perfectly (she'd ignored some of the prescribed format requirements), it received a B instead of an A.

The tragedy was that Anaya internalized this grade as confirmation that she wasn't truly excellent, even though her thinking was exceptional. She started to question herself, to second-guess her unconventional approaches. She began to prioritize fitting into the mold over pursuing her own intellectual curiosity. In trying to protect her grade, she was losing the very thing that made her brilliant.

I remember the first time I called her parents for a conference. They seemed worried that Anaya wasn't getting straight A's. In their mind, if she were capable, she should be at the top of the class. I showed them her project portfolio instead of her grades. I talked about her persistence, her creativity, her willingness to challenge herself and to think in ways that didn't come naturally to everyone. I told them specific stories: the day she spent an entire class period redesigning a solution to a problem because she thought there might be a more elegant approach; the way she'd ask questions that had no single right answer and seemed genuinely interested in exploring the complexity; the moment she realized she'd been wrong about something and how she adjusted her thinking mid-stream without any sense of shame or failure.

For the first time, they saw their daughter not as a set of scores, but as a thinker. The relief on their faces was palpable. They'd been worried because they'd bought into the same system I was trying to help them see beyond. They'd believed that grades were the measure of their daughter's worth and potential. Seeing her portfolio, they understood that Anaya's true education was something far richer than any letter grade could capture.

The Problem With Perfection

Marcus was different. He was bright, but he'd been burned before. In elementary school, he'd been identified as gifted, which came with a particular set of expectations. He had to be right. He had to excel. He had to make it look easy. By the time he reached my classroom, something had shifted. He told me, directly and without much emotion, "If I don't think I can get it perfect, I'd rather not hand it in at all." His grades were low, but his understanding was deep. He could engage in sophisticated discussions, and when he did attempt assignments, his thinking was sound. But the moment he sensed any possibility of error, he'd shut down.

This was perfectionism taken to its logical extreme, and it was paralyzing him. I realized that praising grades alone would only reinforce his belief that performance was what mattered. I needed a different approach entirely.

I started giving credit for process, not just answers. I created assignments where students had to show their thinking, including their mistakes and false starts. I graded based on effort and growth, not just correctness. I celebrated every attempt, every question, every step forward—not with hollow praise, but with specific recognition of what he'd done: "Marcus, I notice you tried a different approach this time. That's growth." "You asked a really thoughtful question that shows you're thinking deeply about this." "You made an error here, and you caught it yourself and corrected it. That's the kind of metacognitive awareness that matters."

His confidence grew. Eventually, his grades did too. But what mattered most was the day he looked at a challenging problem and said, "I'll try." Not "I'll try and probably fail" or "I'll try if you promise to give me credit anyway." Just a simple, direct engagement with something uncertain. He'd moved from seeing risk as catastrophic to seeing it as normal.

Where Real Learning Lives

I started to see that the most important work sometimes happens in the gaps between formal assessments. There's the student who finally speaks up in class, even if what they say is uncertain. There's the group that rallies around a struggling peer, helping them through a tough project. There's the moment when a student, who has always been afraid to fail, takes a risk in front of everyone. There's the breakthrough when a student makes a connection they've never seen before, and you can literally see their face light up with understanding.

These moments don't show up on a report card. You can't put a number on the day a student realizes they're capable of something they thought was beyond them. You can't quantify the growth that happens when a kid who's always been self-conscious about their thinking discovers that their unique perspective adds value to a group discussion.

Sofia was one of my quieter students, the kind who seemed to blend into the background. Her test scores were average, and her homework was turned in but unremarkable. But over the course of the semester, I noticed something. When other students were struggling, Sofia would quietly help them. She'd explain a different way of thinking about a problem and offer encouragement. She never sought credit for it. She seemed to genuinely care about her classmates' understanding. When we did group projects, students actively requested to work with her because they knew she'd push the group forward and make sure everyone's voice was heard.

None of this showed up in her grade. Her GPA was a 3.2, solid, but not exceptional. Yet the impact she had on her classmates' learning and confidence was immeasurable. A grade couldn't capture her value.

I began to share these stories with parents, with students, with colleagues. I wrote notes on report cards: "Sofia showed incredible leadership in her group project." "Malik persisted through a tough week." "Anaya asked a question that changed the way we all saw the lesson." Grades tell one story. The whole student tells another.

Reframing Assessment

What I've come to believe is that we need to fundamentally rethink how we assess and report on learning. Grades should be just one piece of information, not the whole picture. I started experimenting with portfolios, with narrative feedback, with student self-assessment. I asked students to reflect on their own learning: "What do you think you've learned this semester? What evidence shows this learning? Where do you want to grow?"

This shift in how I communicated about learning had surprising effects. Students became more engaged with their own progress. Instead of asking "What did I get?" they started asking "What am I learning?" Parents began to see their children as thinkers and learners, not just as grade earners. And students started to internalize a healthier relationship with success and failure.

I also started to recognize that different students show what they know in different ways. Some are test-takers; others shine in project-based learning. Some are writers; others are presenters or problem-solvers. When I broadened the ways students could demonstrate mastery, I saw previously invisible learners suddenly flourish. A student who struggled with written tests might excel at creating a multimedia presentation. A quiet student might produce stunning written work. A kinesthetic learner might create a model or demonstration that showed sophisticated understanding.

When we look beyond grades, we invite students to see themselves as more than a number. We make space for growth, for risk, and for the kind of learning that lasts a lifetime. We also make space for the reality that learning is not linear, that progress comes in different forms, and that a single snapshot in time cannot capture a student's full potential or trajectory.

The Conversation With Families

One of my biggest challenges was helping families understand why I valued these other forms of assessment so highly. Some parents were skeptical of portfolio-based feedback or narrative comments. They wanted to know their child's GPA, wanted it compared to other students, and wanted the objectivity (or what they perceived as objectivity) of a number.

I learned to meet them where they were, while also expanding their thinking. I'd show them their child's work over time, highlighting growth and development. I'd ask them questions: "What do you notice about how your child's thinking has changed?" "What questions is your child asking now that they weren't asking before?" "In what areas has your child taken more risks or shown more independence?"

Many parents came to appreciate this fuller picture. They began to understand that a student could have a B average and still be a sophisticated thinker, that a student could be getting A's and still be playing it safe, that grades alone told them very little about what their child was actually learning or who they were becoming as a learner.

A System in Need of Change

I recognize that as an individual teacher, I can only do so much to counteract the larger system's obsession with grades. I can't eliminate standardized testing. I can't change college admissions requirements. I can't make society stop treating GPA as the primary measure of a student's worth. But within my classroom, I can create a different culture. I can help my students and their families see grades as one piece of information, not the whole story. I can celebrate the learning that happens outside the traditional assessment structures. I can help students develop a healthier, more growth-oriented relationship with their own learning.

And I can advocate for change. I can speak up in faculty meetings about the need for more authentic assessment. I can work with colleagues to develop alternative ways of tracking and reporting progress. I can help schools think about what we're really trying to measure and whether our current methods actually capture it.

What I know, from years in the classroom, is that when we look beyond grades, we see our students more clearly. We see their strengths that don't fit neatly into a rubric. We see their growth, even when it doesn't translate into perfect scores. We see the whole person—their curiosity, their character, their capacity to think and create and contribute. And when students know they're being seen this way, they often rise to meet that vision.

Reflective Prompts for Educators:

• How can I shift my perspective on assessment to focus more on the learning process than on final grades?

• What steps can I take to create an environment where students feel valued for their effort, creativity, and growth, even if their grades don't fully reflect those qualities?

• How can I provide feedback that highlights not only what students have done well, but also the risks they've taken and the challenges they've overcome?

• What opportunities can I create for students to demonstrate their learning in ways other than traditional tests and assignments?

- How can I help students and their families see the value in alternative forms of assessment, such as portfolios and narrative feedback?

- In what ways can I support students who are perfectionists by praising their efforts and growth, rather than just their results?

- How can I encourage students to take more risks in their learning, knowing that failure and mistakes are integral parts of the process?

- What can I do to help students develop a healthier relationship with grades, one that focuses on learning and growth instead of comparison and competition?

- How can I better communicate the importance of personal growth and intellectual curiosity to families who prioritize GPA and traditional assessments?

- What actions can I take to advocate for changes in the education system to promote more holistic assessments of student progress and potential?

Chapter 4
The Language Barrier

There's a particular kind of bravery that comes with learning in a new language. Every day is an act of translation of ideas, of feelings, of self. It's exhausting, humbling, and, at times, isolating. Language learners in our classrooms are performing an extraordinary cognitive feat that native speakers rarely appreciate. They're not just learning new vocabulary or grammar rules. They're learning to think in a new way, to express themselves through unfamiliar structures, to navigate social interactions where the unspoken rules are different from what they've known.

Ahmed entered my classroom halfway through the year. His family had moved across continents, and English was still new to him. He came from an affluent background, had been educated in prestigious schools, and was, by any measure, an intelligent and capable student. Yet in my classroom, he was rendered nearly speechless. In the first few weeks, he barely spoke. When I called on him, he'd look at me with quiet panic, unsure whether to answer or retreat. His eyes would dart around the room, as if searching for an escape route. I could see the internal struggle: he knew the answer, I could tell by his expression, but the thought of speaking English in front of his peers seemed to paralyze him.

What struck me most was the visible effort it took for him just to be present. He'd arrive each day looking drained, as if the simple act of navigating a classroom conducted entirely in a foreign language had exhausted him. But I also noticed something remarkable: he watched everything. The way I moved around the room, the way his classmates argued over answers, the way the chalk squeaked against the board. He was learning, even if he couldn't always show it. His silence wasn't a sign of passivity or lack of engagement. It was the silence of intense concentration, of someone constantly absorbing and processing.

I remember thinking about what it must be like to sit in a classroom where everyone around you is communicating fluently, making jokes you don't quite understand, and engaging in debates with ease. Imagine being surrounded by conversation but unable to fully participate. Imagine knowing what you want to say but not having the words, or having the words but not being confident enough to speak them. Imagine the fear that comes with every raised hand, the knowledge that you might stumble over pronunciation, use the wrong tense, or say something that makes you sound less intelligent than you actually are.

Creating Space for Language Learners

I started making deliberate changes to my classroom to support Ahmed and the other language learners who were struggling. I started using more visual pictures, diagrams, and gestures. I realized that much of what I was teaching could be understood through images and movement, not just through words. When I was teaching about photosynthesis, I didn't just talk about it; I drew it, acted it out, and showed videos. When I was explaining a concept, I used my hands, my body, and my face to communicate meaning. This wasn't just helpful for language learners; it actually made the content more accessible and memorable for everyone.

I let Ahmed pair up with a classmate who spoke his home language. This was transformative. Having the option to explain his thinking in Arabic first, then translate to English, seemed to unlock something in him. He could work through his ideas more fully and then translate them into English when he felt ready. This wasn't cheating or shirking; it was a legitimate learning strategy that allowed him to access higher-level thinking while he continued to develop his English proficiency.

I gave him extra time on assignments. He didn't get different assignments or lower standards—just more time to process, to think, to find the words. I invited him to stay after class, and we'd sit together while I'd explain something he'd missed or wasn't sure about. These conversations happened in a low-pressure environment, one-on-one, where he could ask questions without worrying about how it looked in front of his peers.

And I celebrated every effort, no matter how small. I didn't wait for perfect English or correct answers. I celebrated the attempt. When Ahmed raised his hand tentatively to answer a question, even if his response was halting or incomplete, I acknowledged it. "That's a good start, Ahmed. Can you tell me more?" I was signaling that communication was valued over perfection, effort over flawlessness.

One day, during a science experiment, something shifted. We were observing chemical reactions, and Ahmed blurted out an answer—hesitant, with a slight accent, but correct. He'd noticed something the other students had missed. The class cheered. I could see the pride on his face. It was as if a weight had lifted. From that day on, his hand went up a little more often. He still spoke carefully, still sometimes searched for words, but he was no longer afraid.

Understanding the Complexity of Language Learning

What I've come to understand is that language barriers are rarely just about vocabulary. Sometimes, it's about confidence. Sometimes it's about missing the cultural cues that everyone else takes for granted. American classrooms operate on specific, often unspoken norms. Students are expected to speak up, to ask questions, and to challenge ideas respectfully. They're expected to be comfortable with ambiguity and with not knowing answers. They're expected to take intellectual risks. But in many other educational systems, these norms are different. Students are expected to listen quietly, to absorb information, to speak only when called upon, and to defer to authority. For a student coming from that background, the implicit invitation to participate in American classroom discussion can be confusing or even alarming.

I learned to be explicit about expectations: "It's okay to make mistakes here." "We want to hear your ideas." "There's no such thing as a stupid question," I said these things repeatedly, and I modeled them. When I made a mistake, and I made sure to make them visible, I'd acknowledge it: "Oops, I made an error. Let me think about that again." I'd ask genuine questions when I wasn't sure about something. I'd admit confusion. I was trying to create a culture where imperfection was normal.

I encouraged group work, where students could practice language in lower-stakes settings. Instead of putting Ahmed on the spot in front of the whole class, I'd have students work in small groups. He could practice his ideas with two or three peers before sharing with the larger group. He could hear other students working through their thinking, could see that making mistakes was part of the process, and could build confidence gradually.

Mei was another student who taught me about the complexity of language barriers. She came from a school where speaking up was discouraged. In her previous educational context, students who asked too many questions or offered opinions were seen as disrespectful or showing off. She was brilliant, her written work was sophisticated and thoughtful, but she'd been taught that silence was a virtue. She was fluent in English, so technically, she didn't have a language barrier. But she had a cultural barrier, an internalized belief that her voice didn't belong in the classroom.

I watched her sit quietly through discussions, taking careful notes, clearly following the content but never participating. I tried calling on her directly, but she'd give short, minimal answers and then retreat. I realized that pushing her to participate publicly was only reinforcing her belief that it was wrong to do so.

Instead, I gave her roles that played to her strengths: note-taker, researcher, and discussion leader in small groups. I started asking her to prepare written reflections that we'd discuss one-on-one. I invited her to email me her thoughts on the day's lesson. Slowly, she found her voice. By the middle of the year, she was volunteering comments in class, though still more reserved than some students. By the end of the year, she was leading a small group discussion with confidence and clarity.

Cultural Wealth in the Classroom

I realized that my classroom was richer for the diversity of languages and perspectives. When I made space for students to share their home cultures through stories, music, or traditions, the entire class learned. Ahmed brought in Arabic poetry and explained its structure and rhythm. Mei shared traditional Chinese calligraphy and taught us about the philosophy behind it. Other students shared foods from their families' countries of origin, and we'd have conversations about how food connects to culture and identity.

We celebrated holidays from around the world. We didn't just acknowledge them; we explored them. We read stories, watched videos, listened to music, and tried foods. And crucially, we centered the students from those cultures as the experts. Ahmed wasn't just a student in my class; he was a guide to Arabic culture and language. Mei wasn't just taking a test; she was sharing wisdom from her heritage.

This shift changed the entire dynamic of the classroom. Suddenly, speaking another language wasn't a deficit; it was an asset. Having a different cultural background wasn't something to apologize for; it was something to celebrate. Students who had felt like outsiders because of their language or cultural differences started to see themselves as valuable contributors to the classroom community.

I also honored the different ways of seeing and being that these students brought with them. Some students approached problems sequentially, step by step. Others saw patterns and connections intuitively. Some valued individual achievement; others prioritized group harmony. Instead of privileging one way of thinking, I made space for multiple approaches. I asked students to explain their thinking, to describe why they solved a problem the way they did, to articulate their cultural values around learning and achievement.

The Invisible Labor of Language Learning

What I've come to deeply appreciate is the invisible labor that language learners engage in every single day. It's not just the cognitive effort of processing language. It's also the emotional effort. Language learners often experience anxiety, frustration, and a sense of inadequacy that native speakers rarely feel. They might feel embarrassed about their accent or their grammar. They might worry that they're being judged. They might feel isolated, unable to access the social connections that happen through language.

I started trying to imagine what it would feel like to be in Ahmed's shoes. To sit in a class for forty-five minutes and not fully understand everything being said. To know that you're missing jokes and references that everyone else seems to find funny. To have something important to contribute but not have the words to express it fluently. To hear yourself speak and know that you sound different, less capable, less intelligent than you actually are.

This empathy led me to make different choices in my classroom. I slowed down. I didn't feel like I had to cover as much material at a rapid pace. I left space for processing. I repeated information in different ways. I used written materials, not just verbal instruction. I gave students time to formulate their thoughts before calling on them.

I also started paying attention to the moments when language learners struggled. Instead of moving on quickly or covering for them, I'd pause and give them time. "That's a great question, and I want to give you time to think about it. Take your time." This signaled that their contribution was valuable enough to wait for, that I wasn't just going to fill the silence with my own voice.

Beyond the Classroom

I realized that supporting language learners extended beyond classroom strategies. It also meant advocating for them in other contexts. When I heard a teacher or administrator make a dismissive comment about a student's language proficiency, I'd gently push back. "Ahmed's English is developing, but he's incredibly thoughtful and intelligent." I'd make sure that language learners were included in advanced or accelerated programs, not just placed in remedial tracks. I'd advocate for ESL support and resources.

I also made sure that language learners knew they could come to me with concerns or difficulties. I'd check in regularly: "How are you feeling about the class? Is there anything I can do to help?" These conversations sometimes revealed barriers I hadn't anticipated. One student shared that she felt self-conscious about her accent and had been avoiding speaking. Another shared that he was embarrassed about needing extra time and would prefer not to be visible about it. I used this feedback to adjust my approach.

I also connected language learners with peer tutors, not to remediate but to build relationships and provide additional practice. I found that peer relationships were often more powerful than teacher-led instruction. When Ahmed started working with a native English-speaking peer who was patient and genuinely interested in his perspective, his confidence grew more than it had from my individual check-ins.

The Transformation

By the end of the year, Ahmed's transformation wasn't about fluent, perfect English. It was about belonging. He spoke up in class. He contributed to discussions. He asked questions. He made friends. He participated in group projects with confidence. He still had an accent. His grammar wasn't perfect. But he was no longer paralyzed by fear or self-consciousness.

More importantly, his presence changed the classroom. Other students developed patience and understanding for language learners. They learned to listen carefully, to give time for processing, and to value communication over perfection. They heard Ahmed's ideas and recognized his intelligence. The classroom became more inclusive for everyone, not just for language learners.

Reflective Prompts for Educators:

- How can I create a classroom environment where language learners feel safe to express themselves without the fear of making mistakes?
- What steps can I take to better understand the emotional and cognitive effort language learners are making every day in my classroom?
- How can I use visuals, gestures, and body language to support students who are learning in a second language?
- What can I do to help students like Ahmed feel empowered to participate, even when they're still developing language skills?
- How can I celebrate small victories and efforts in language learners, showing them that progress is more important than perfection?
- What strategies can I implement to encourage students to view their language differences as strengths rather than barriers?

- How can I support language learners' cultural backgrounds and make space for their perspectives in classroom discussions?
- What can I do to reduce the invisible labor of language learning, so that students don’t feel isolated or misunderstood in class?
- How can I shift my classroom culture to value communication and effort over flawless grammar or pronunciation?
- What are some ways I can advocate for language learners outside the classroom to ensure they’re supported in all areas of their education?

Chapter 5
Lessons in Life

Teaching is supposed to be about imparting knowledge, but the longer I do it, the more I realize how much I'm learning from my students. The deepest lessons aren't about math or science—they're about resilience, empathy, and the quiet courage it takes to keep showing up. In fact, some of the most profound education I've received has happened not when I was standing at the front of the classroom delivering content, but when I was paying attention to the stories unfolding within my classroom walls.

The Perseverance of Jamal

Jamal taught me about perseverance in a way that no professional development workshop ever could. Every day, he balanced school with responsibilities most adults would find overwhelming. He cooked dinner for his family most nights. He babysat his younger siblings after school. He helped with household chores and supported his mother in navigating systems and paperwork that she didn't fully understand. He worked a part-time job on weekends. And yet, he still made it to class sometimes late, always tired, but present.

I'd watch him struggle through assignments. His work was often incomplete or submitted late. But what I noticed, if I looked beyond the grade, was the growth. He was learning. He was trying. He was doing the best he could with the time and energy he had available. When he finally finished a challenging project, a deep dive into a topic of his choice, he looked at me with a mixture of pride and exhaustion and said, "I didn't think I could do it, but I tried anyway."

Those words have stayed with me. I carry them with me on my hard days, when I'm wondering if I'm making a difference, if my efforts matter. "I didn't think I could do it, but I tried anyway." That's not the definition of success that we often celebrate. We celebrate the straight A's, the perfect projects, the students who make it look easy. But this is the act of trying despite doubt, despite exhaustion, despite overwhelming odds, this is the real definition of success.

Jamal also taught me something about fairness and systems. He wasn't failing because he lacked ability or motivation. He was struggling because the system wasn't built for students like him. The expectation that all students would have the same amount of time and support outside of school is an expectation that ignores the reality of many students' lives. When I became aware of this, I tried to adjust. I gave Jamal extensions without penalty. I broke assignments into smaller pieces. I acknowledged that he didn't need to be "fixed," the system did.

The Creativity of Maria

Maria showed me the power of perspective and the danger of rigid thinking. She was a good student, not top of the class, but solid. But what made her special was her ability to see problems from angles that most people wouldn't consider. In one class, when I was teaching a standardized algorithm for solving a particular type of math problem, she raised her hand and asked, "But what if you did it this way instead?" She explained an approach that was unorthodox, that broke the "rules" I'd taught, but that worked. In fact, it was more elegant than the standard method in some ways.

My first instinct was to correct her, to redirect her back to the "right" way. But I paused. I looked at her work more carefully. It was correct. It was actually quite clever. And I realized, in that moment, that I had as much to learn from her as she did from me. Her creativity pushed me to rethink my own assumptions about the "right" way to solve a problem.

This became a turning point in how I taught. Instead of presenting information as fixed and final, I started positioning it as one way of thinking about a problem, not the only way. I invited students to find alternative approaches. I celebrated unconventional solutions. And I noticed that students who had previously seemed unmotivated became engaged when they realized that their unique perspectives were valued.

Maria taught me that intelligence isn't just about absorbing and reproducing information. It's about synthesis, creativity, and the ability to see connections that others miss. It's about having the courage to propose a different way when everyone around you is doing something the standard way.

The Anxiety of Elias

Elias struggled with anxiety, sometimes paralyzed by the fear of failing. For him, a challenging problem or an unexpected question could trigger a physiological response: racing heart, difficulty breathing, sometimes even freezing up completely. His anxiety wasn't a character flaw or something he should just "get over." It was a real barrier to his learning, one that required patience and specific strategies to navigate.

I learned to break tasks into manageable steps for Elias, to check in frequently, and to celebrate every tiny victory. When he completed an assignment that he'd been anxious about, his pride was immense. He wasn't just proud of the work itself; he was proud of having done something that scared him. I learned to recognize and honor that. "I see that this was challenging for you, and you did it anyway. That's what courage looks like."

Elias taught me about the importance of encouragement and of noticing progress. He showed me that sometimes, the most important thing I can do as a teacher isn't to deliver content or assess knowledge. It's to believe in students even when they doubt themselves. It's important to notice their progress, even when it's small. It's to create an environment where trying, even when you might fail, is celebrated.

I also learned from Elias about the limits of my own ability to fix everything. His anxiety was deep and persistent, and no amount of my encouragement could completely resolve it. But what I could do was create a safe space within my classroom, connect him with professional support, and consistently communicate that his worth wasn't dependent on his academic performance or his anxiety levels.

The Kindness of Sofia

Sofia, despite her academic struggles, was a model of kindness. She wasn't the top student in the class. She didn't always turn in homework on time. Her test scores were average. But she cheered for her classmates. She offered help freely, without being asked. She never hesitated to give a word of encouragement. When another student was struggling, Sofia seemed to have a sixth sense for it. She'd sit next to them, ask if they needed help, and offer to work through the problem together.

Her compassion was remarkable, and it was a reminder that character matters as much as achievement. In a system that often measures students by their grades and test scores, Sofia would be considered average. But by any measure of human goodness, she was exceptional. She made the classroom a better place simply by showing up and caring about her peers.

Sofia taught me to expand my definition of success. She taught me that not all important learning can be measured by a test. She taught me that the students who struggle academically might have tremendous gifts and wisdom to offer. And she taught me that kindness is a form of intelligence that our educational system rarely acknowledges or rewards.

I also started thinking differently about how I could support Sofia's learning. Instead of focusing on her academic deficits, I started building on her strengths. Her empathy and social awareness were gifts. I found ways for her to use those gifts in the classroom—leading group discussions, mentoring struggling peers, and helping to build community. And as her confidence grew, her academic performance improved as well.

Learning Is Reciprocal

Every year, students teach me that what matters most isn't what's on the test, but what's in their hearts. They show me that patience and kindness are the real curriculum, and that learning is a two-way street. When I approach my teaching with genuine curiosity about my students' lives and perspectives, when I'm willing to learn from them as they learn from me, the classroom becomes a different kind of space.

I've learned about resilience from students who keep showing up despite overwhelming challenges. I've learned about creativity from students who see problems differently than I do. I've learned about anxiety from students who struggle with their own fear and yet still participate. I've learned about kindness from students who care more about their peers' well-being than their own grades.

These aren't lessons I could have learned from a textbook or from a professional development workshop. They're lessons that come from paying attention, from being in relationship with students, from being willing to be changed by them.

The classroom is a place of shared growth. I am a better teacher and a better person because of the lessons my students have given me. I'm more patient. I'm more flexible. I'm more aware of the systems and structures that privilege some students and disadvantage others. I'm more willing to admit when I'm wrong or when I don't know something. I'm more attuned to the quiet strengths that don't always show up on report cards.

Reflective Prompts for Educators:

- How can I recognize and honor the quiet moments of courage in my students, especially when they're pushing through fear or doubt?

- What steps can I take to create a classroom environment where kindness, empathy, and emotional intelligence are celebrated just as much as academic achievement?

- How can I encourage students like Jamal, who face significant outside challenges, to believe in their growth and perseverance, even when their academic progress is slower?

- What strategies can I use to foster creativity in my students and encourage them to think outside the box and challenge the "rules" in constructive ways?

- How can I create more opportunities for students with anxiety, like Elias, to succeed by breaking tasks into manageable pieces and celebrating each step forward?

- In what ways can I shift my definition of success to include qualities like perseverance, empathy, and creativity, rather than focusing solely on grades and test scores?

- How can I build on students' strengths, like Sofia's compassion and kindness, to help them feel valued and empowered in the classroom?

- What can I do to create a reciprocal learning environment where I learn from my students as much as they learn from me?

- How can I better support my students by becoming more aware of the systems that affect their success and finding ways to adjust my teaching to meet their unique needs?

- How can I encourage my students to see learning as a process of growth and not just a series of tests or benchmarks to meet?

Chapter 6
Overcoming the Odds

Every classroom has its quiet heroes, students who battle challenges most adults never see, who keep showing up even when the world seems determined to hold them back. Their struggles are often invisible: unstable housing, food insecurity, family trauma, illness, untreated mental health conditions, or the slow, grinding pressure of poverty. For these kids, school isn't just a place to learn. It's often the most stable, predictable part of their lives. It's a place where someone cares about them, where there's consistency, where they have a chance to envision a different future for themselves.

These are the students who deserve our deepest attention and most thoughtful support. Yet they're often the ones who are most overlooked or underestimated. Their struggles can be interpreted as laziness or indifference when the reality is usually far more complex. They're juggling things that would break most of us, and they're still showing up to class.

Kayla's Journey

Kayla came to my class mid-year, trailing missed assignments and a cloud of sadness. Her parents had separated, and she'd moved twice in three months. Her mother was struggling financially and emotionally. Her father was largely absent. Kayla was caught in the middle, worried about her mother, uncertain about what came next, grieving the loss of stability.

At first, she barely spoke. She'd come to class, sit in the back, and seem to disappear into herself. Her assignments were incomplete or missing. When I'd ask her about them, she'd make excuses, but I could sense that something deeper was going on. She wasn't lazy or unmotivated. She was scared and overwhelmed.

Instead of focusing on grades, I started focusing on connection. I'd ask her about her day. I'd check in about how she was doing, not just academically. I'd make time to listen, really listen to what she was saying and what she wasn't saying. Slowly, through essays and small projects, she began to share her story. She started writing about her feelings, her fears, her confusion. Her writing was raw and honest.

One day, she asked if she could read aloud a piece she'd written about finding hope when everything felt hopeless. I said yes, not knowing what to expect. As she stood in front of the class, her voice trembled. She was nervous, vulnerable, exposing her pain to a room full of peers. But she finished. When she was done, the class applauded her not for the grade, not for the quality of the writing, but for her courage. I could see the relief on her face. She'd been seen. She'd been heard.

That moment became a turning point for Kayla. She continued to struggle with the practical realities of her situation, unstable housing, and financial insecurity, but she had a space where she could process those struggles. She had an adult who acknowledged her strength and her resilience. She had evidence that her story mattered, that her voice had value.

By the end of the year, Kayla's grades had improved, but more importantly, she'd found a sense of agency. She wasn't just a victim of her circumstances; she was someone who was navigating them, who was surviving, who was still learning and growing despite everything.

Malik's Determination

Malik's attendance was spotty, his grades low. He'd miss weeks of school at a time, and when he returned, he'd be so far behind that catching up felt impossible. I almost gave up on him. It would have been easier, in some ways, to just let him slip through the cracks. He was so far behind, and the system isn't designed to make it easy for kids like Malik to catch up.

But then I had a conversation with him. I asked, directly, "Do you want to catch up? Or have you decided it's not worth it?" He surprised me by saying that he did want to catch up. He did want to improve his grades. He did want to graduate. He just didn't know how, and he felt so far behind that it seemed impossible.

Together, we made a plan. Not a massive overhaul, which would have been overwhelming, but small steps. We identified the most important things he needed to catch up on. We made a schedule that was realistic, given his attendance challenges. We checked in regularly, not just about assignments, but about what was getting in the way of attendance. Some days he'd stay after school with me. Some days we'd do work during lunch. We celebrated every step forward, no matter how small.

Progress was slow, but it was steady. There was a day when he turned in an assignment that was incomplete but showed effort. There was a day when he attended school for a whole week straight. There was a day when he passed a quiz. None of these things would seem remarkable to a student who doesn't struggle with attendance. But for Malik, they were monumental.

By the end of the year, Malik had not only improved his grades but had become an example for others. Other students saw him struggling and then persisting, and they believed that change was possible. He showed what's possible when you refuse to give up, when you have someone who believes in you and supports you, when you take it one step at a time instead of trying to fix everything at once.

Understanding Invisible Barriers

What I've learned through students like Kayla and Malik is that the barriers to success aren't always academic. Sometimes, they're logistical. How do you focus on homework when you don't know where you'll sleep? How do you complete a project when you don't have reliable internet access? How do you keep trying when every system seems designed to keep you down?

I've also learned that these barriers are often invisible to those of us who haven't experienced them. A student misses an assignment, and we assume it's because they didn't prioritize it. In reality, they might have been dealing with a housing crisis. A student seems disengaged, and we assume they don't care. In reality, they might be dealing with depression, grief, or family trauma.

As teachers, I've come to realize that we can't solve all of these problems. We can't provide housing, food, or mental health treatment. But what we can do is acknowledge the realities our students face. We can be flexible. We can offer grace. We can celebrate persistence. We can refuse to give up on students, even when the odds seem overwhelming.

Building Resilience

These aren't stories of overnight transformation. They're stories of persistence of one small step after another, of teachers and parents and students working together, celebrating every inch of progress. What I've learned is that resilience is built, not given. It doesn't spring from nowhere. It comes from being seen, being supported, and being believed in.

When Kayla shared her story with the class and saw the response, she was building resilience. When Malik tock one small step toward catching up, and someone acknowledged it, he was building resilience. Resilience isn't about toughing it out or pulling yourself up by your bootstraps. It's about having relationships, having support systems, having people who believe in you even when you don't believe in yourself.

As teachers, we can be a crucial part of those support systems. We can be the person who says, "I see what you're dealing with, and it's real. And I also see your strength. And I'm going to help you." We can be the person who refuses to give up, who keeps offering support even when the student pushes back, who celebrates progress in its smallest form.

The Ripple Effect

As teachers, we can't fix every problem. But we can offer consistency, patience, and hope. Sometimes, that's enough to help a student reach their next milestone. The ripple effect is real: when one student overcomes the odds, others start to believe they can, too.

I've seen this happen in my classroom. Malik's persistence inspired other struggling students. Kayla's vulnerability opened the door for other students to share their stories. When one student sees a peer overcome significant obstacles and succeed, something shifts. The narrative changes from "it's impossible" to "it's possible, and here's what it looks like."

I've also learned that supporting students who are overcoming the odds requires thinking beyond the individual student. It requires collaboration. It requires connecting with counselors, social workers, and community organizations. It requires advocating for systemic changes that would make it easier for these students to succeed instead of just offering individual support as a band-aid.

It requires believing that these students deserve more than just survival. They deserve to thrive. They deserve to have dreams. They deserve to have adults in their lives who believe those dreams are possible.

Reflective Prompts for Educators and Parents:

- How can I recognize and address the invisible barriers that might be preventing my students from fully succeeding, especially when those barriers are not immediately obvious?

- What steps can I take to ensure that students like Kayla and Malik, who face significant challenges outside of school, feel supported and valued in the classroom?

- How can I create a classroom culture where resilience is celebrated, not just academic success, and where every student feels acknowledged for their efforts, no matter how small?

- What can I do to offer grace and flexibility to students who are struggling with life circumstances that impact their ability to keep up with schoolwork?

- How can I build stronger relationships with my students to foster trust and support, especially for those who may feel isolated or disconnected?

- What role can I play in helping my students develop not just academic skills, but personal resilience, empathy, and the courage to keep going despite setbacks?

- How can I ensure that every student in my class, regardless of their circumstances, has a voice and is seen as a valuable contributor?

- What are some practical ways to create an inclusive classroom environment that recognizes the diverse realities students face and supports their holistic growth?

- How can I collaborate with other professionals, like counselors and social workers, to create a network of support for students dealing with systemic challenges?

- What changes can I advocate for in the education system to better support students who face significant challenges outside of school, so they have a better chance to thrive?

Chapter 7
The Hidden Struggles of Being a Teacher

From the outside, teaching is often romanticized as endless inspiration, summers off, and the joy of shaping young minds. There's a cultural narrative about teachers being dedicated heroes, martyrs even, willing to sacrifice everything for their students. But the reality is more complicated, more human, and far more difficult than that narrative suggests. Behind every lesson plan is a human being juggling a hundred unseen tasks, wrestling with doubts, and carrying the emotional weight of dozens of young lives. Behind the classroom door, teachers are managing not just content delivery, but the emotional complexities of adolescence, the structural inequities of education, the pressure of standardized testing, and the persistent question of whether any of it is actually making a difference.

The Relentless Exhaustion

There are days when I arrive at school already exhausted. Nights spent grading papers—stacks of them, each one representing a student's effort that deserves careful feedback, not a quick skim. Drafting emails to parents, explaining why their child is struggling, or celebrating their progress. Worrying about the quiet kid in the back row who hasn't spoken in weeks and seems to be retreating further into themselves. Replaying conversations with students, wondering if I said the right thing, whether I missed an opportunity to make a difference.

I'll be honest: there are mornings when I sit in my car in the parking lot for ten minutes before going in, just trying to muster the energy to face the day. On those mornings, the thought of being upbeat and supportive for seven hours feels impossible. Yet the moment I walk into my classroom, I have to switch modes. I have to be "on" present, energetic, and available. My students need that from me. They don't need to see my exhaustion or my doubts.

The pressure to be present, upbeat, and supportive every minute is relentless. There's no "off" switch—no way to leave your work at the door. Your phone buzzes with an email from a parent in the evening. You see a former student's post on social media and wonder how they're doing. You're at the grocery store and run into a student's sibling, and suddenly you're having a work conversation when you were just trying to buy milk. Teaching doesn't have boundaries the way other professions do. It seeps into your personal time, your mental space, your sense of self.

The physical exhaustion is real, but the emotional exhaustion is deeper. Teachers absorb more than they show. A student's pain becomes your own. If a student shares that they're struggling with depression, you carry that. If a student reveals that they're being hurt at home, you carry that. If a student comes to class with red-rimmed eyes and won't tell you what's wrong, you carry the not-knowing as well. Their struggles linger in your mind long after the bell rings. You celebrate their victories, but you also grieve their losses, replaying conversations and wondering if you could have done more.

I have a student whose parents are divorcing, and she confided in me about how scared she is. Now, when I see her, I'm hyperaware of her pain, even when she's trying to hide it. I have a student who's been bullied, and I find myself checking in on him more than I do with other students, worried about his well-being. I have a student with an undiagnosed learning disability who's being blamed by his parents for "not trying hard enough," and my heart breaks for him.

These students become part of my internal landscape. I think about them when I'm not at school. I research strategies to help them. I worry about their futures. And while this investment matters—it's part of what makes a good teacher, it also comes at a cost to my own wellbeing.

The Impossible Balancing Act

Work-life balance is a myth most weeks. Evenings disappear into lesson planning. I'll start at 6 PM, thinking I'll just get one thing done, and suddenly it's 11 PM, and I'm still at it. Weekends vanish under piles of grading. I've graded papers on planes, in restaurants, and on family vacations. There's always more to do. Always more students who deserve careful feedback. Always another unit to plan, another assessment to create, another email to answer.

For teachers with families, the guilt cuts both ways, never quite enough at work, never quite enough at home. I miss my kids' soccer games because of grading. I'm mentally at school during family dinner, thinking about a lesson I need to revise or a student I'm worried about. My partner asks me a question, and I have to ask them to repeat it because I wasn't really listening; I was thinking about something from class.

And then there's the guilt at school. I could be doing more to support my students. I could be creating more engaging lessons. I could be differentiating more, reaching more learners, and giving more individual attention. The needs are infinite, and my time and energy are finite. Something always has to give, and that something is often my own well-being.

I know teachers who've had panic attacks while grading. I know teachers who've cried in their cars after particularly difficult days. I know teachers who've taken medication for anxiety and depression. I know teachers who've left the profession because it was simply unsustainable. And I know that if we talk about this at all, it's often in whispered conversations in the staff room, as if admitting our struggles is unprofessional or suggests we're not cut out for teaching.

The Emotional Labor

The emotional labor is real and often underestimated. Supporting kids with trauma requires showing up for them emotionally while maintaining appropriate boundaries. Supporting kids with learning differences means problem-solving constantly, trying new strategies, and adapting on the fly. Supporting kids with language barriers means being patient with communication, celebrating little progress, and advocating within the system. Supporting kids with anxiety means creating safe spaces, validating their fears, and helping them develop coping strategies. And supporting kids with family problems means holding space for their pain while not having the professional training to actually treat the underlying issues.

This emotional labor is deeply, profoundly rewarding. But it's also draining. Teachers give, and give, and give. Too often, they forget to refill their own well. There's an implicit expectation that teachers are endlessly available, infinitely patient, and completely self-sacrificing. We internalize this. We believe it ourselves. We stay late to help a student, sacrifice our lunch to work with a struggling peer, and spend our own money on classroom supplies. And while these acts of generosity are beautiful, they're also unsustainable.

I've had to learn, over the years, that saying no is sometimes an act of survival. I can't stay late every day. I can't help every student with every problem. I can't be the sole source of support for kids who are dealing with serious issues that require professional intervention. And when I finally started saying these things out loud, to myself, to others, it felt selfish. But it wasn't. It was necessary.

The Constant Self-Doubt

There are weeks I question if I'm enough, if I'm making a difference, if my efforts matter. The expectations, from parents, administrators, society, and myself, are sky-high. Parents expect teachers to educate their children, discipline them, teach them values, keep them safe, address their learning differences, and somehow do all of this while also being sensitive to family circumstances that we don't fully understand. Administrators expect test scores to improve, behavior to be managed, paperwork to be completed, and innovation to happen simultaneously. Society expects teachers to solve problems that stem from systemic inequity, poverty, trauma, and inadequate mental health services. And I expect all of this from myself, plus I expect to inspire my students, to make learning engaging, to see every student, and to meet every student where they are.

Failure, even imagined, feels personal. When a student doesn't do well on a test, I wonder what I could have taught differently. When a student drops out, I replay every interaction, wondering if I missed something, if I could have prevented it. When a student tells me they hate my class, I take it to heart. When a parent criticizes my teaching, I spiral into self-doubt. The line between professional feedback and personal attack feels blurry, and I often collapse them.

I've had to consciously work on separating my self-worth from my students' performance. I'm not responsible for whether every student passes or fails. I can't force students to care about learning. I can't fix every problem a student brings into my classroom. I can provide good instruction, create a supportive environment, and show up with care and intention. And then I have to accept that the outcome is not entirely within my control.

This has been one of the hardest lessons for me to learn.

The Joy That Keeps Us Going

But there's also joy: the relationships, the breakthroughs, the knowledge that you matter to someone. Those moments keep you going. The day a student's face lights up when they finally understand something they've been struggling with. The moment a formerly silent student speaks up in class. The email from a former student, years later, saying that something you taught them changed their life. The conversation with a student where they share something real and vulnerable, and you get to bear witness to their growth.

These moments are the fuel. They're the reason we do this. And they're often small enough to miss if you're not paying attention. A quiet "thank you" from a student. A parent mentions that their child talks about your class at home. A piece of work that shows real growth. These are the moments that make the exhaustion, the frustration, the self-doubt almost worth it.

I've learned that self-care isn't selfish, it's survival. I started setting boundaries. I decided that I wouldn't grade papers after 8 PM. I took weekends where I completely separated from work. I said no to some committees and extra responsibilities. I made time for friendships, for exercise, for hobbies that had nothing to do with teaching. I started being honest about what I could and couldn't do. "I can't attend that meeting because I need to pick up my kid." "I can't take on another responsibility right now because I'm at capacity." "I'm struggling, and I need support."

These felt radical, almost forbidden. But they were necessary. And I've noticed that when I take care of myself, I'm actually a better teacher. I have more patience. I'm more present. I'm less likely to snap at a student or to make decisions from a place of exhaustion. My students benefit when I'm well.

I've also learned to celebrate small wins. Not every day is going to be a breakthrough. Some days, the win is just that I showed up. Or that I didn't lose my temper. Or that I helped one student instead of feeling bad that I couldn't help everyone. Or that I made time to eat lunch instead of working through it. These are victories worth acknowledging.

The Importance of Community

Teachers are human. We need support, too. Professional communities, understanding administrators, and a kind word from a colleague matter. The best schools are the ones where teachers are seen, heard, and cared for, just like the students.

I've been fortunate to work in schools where there's a strong teacher community. Where can we talk openly about our struggles? Where we support each other, not just professionally but emotionally. Where a colleague notices that you're having a hard day and sits with you during lunch. Where we laugh together, complain together, and celebrate together.

I've also worked in schools where it's every teacher for themselves, where talking about struggles is seen as weakness, and where there's competition instead of collaboration. In those environments, the exhaustion is worse. The isolation compounds the difficulty. Teaching feels lonely and hopeless.

The difference that administrator support makes cannot be overstated. An administrator who sees their teachers, who acknowledges the difficulty of the work, who advocates for reasonable class sizes and planning time, who supports professional development, who treats teachers with respect—that makes a profound difference. It signals that your well-being matters, not just your productivity. It creates a culture where you're not expected to sacrifice everything.

I've also learned the power of peer support. Having other teachers to process with, to brainstorm with, to complain to (because sometimes you just need to vent), to celebrate with, this sustains me. We share strategies, we normalize struggle, we remind each other that we're not alone in this.

Moving Forward

Teaching is hard. Harder than most people understand. It requires emotional intelligence, endless flexibility, deep subject matter knowledge, pedagogical skill, patience, creativity, and the ability to see potential in people who might not see it in themselves. It requires showing up on days when you're sick or sad or exhausted. It requires caring deeply about people who will forget you in five years. It requires constant growth and change in a system that often resists both.

And it's also one of the most meaningful things I've ever done. Because when I manage to see a student clearly, when I can offer support that actually helps, when I'm part of a learning community that's genuinely trying to grow and change, that matters. That's enough.

I'm learning to hold both truths simultaneously: that teaching is impossibly hard, and that it's worth doing. That I'm giving everything I have, and that it's still not quite enough. I need to take care of myself, and that part of me will always sacrifice for my students. That I'm human and flawed and limited, and that I still have the power to make a difference in someone's life.

Reflective Prompts for Educators:

- How can I set bold boundaries without guilt, ensuring I stay energized and fully present for my students?
- What steps can I take today to give my students the emotional support they need while also protecting my own well-being?
- How can I shift my focus to celebrate every small win, redefining success to fuel my passion for teaching?
- What can I do to make emotional resilience as celebrated and valued in my classroom as academic achievement?

- How can I create a powerful support network with colleagues to share, inspire, and elevate each other every day?
- How can I meet my students' emotional needs while still keeping my energy high and my boundaries intact?
- What bold actions can I take to advocate for the changes needed to make teaching sustainable and effective for all?
- How can I empower myself to see my students' strengths, offer support, and still leave work at school without carrying it home?
- What can I do today to create a school culture that celebrates teacher well-being and fosters a thriving, supportive community?
- How can I embrace the challenges of teaching with excitement and purpose, knowing it's tough but ultimately worth every single moment?

Chapter 8
How to Raise Every Hand

Every teacher pictures a classroom where every hand is in the air, every voice eager to join in. The reality is messier. Some kids are born ready to participate; others need time, space, and a different kind of invitation. The truth is that participation is deeply connected to identity, safety, personality, and prior experiences. It's not just about academic readiness. It's about whether a student feels like their voice belongs in the room.

Understanding the Barriers to Participation

Luis was the kind of student who melted into the background. He was intelligent, his written work showed sophisticated thinking, but he was almost invisible in classroom discussions. Speaking up in class terrified him. At home, he'd learned that silence was safer than risking a mistake. His father was highly critical. When Luis made errors, they were pointed out harshly. When he succeeded, it was expected, not celebrated. He'd internalized the message that it was better to say nothing than to say something wrong.

I watched him during discussions, and I could see him mentally working through problems. He'd write notes to himself. He'd mouth the answer silently. But when it came time to actually speak, he'd retreat. For weeks, I tried to draw him out. I'd ask him direct questions, hoping that putting him on the spot gently would help. It didn't. It made him more anxious. I'd praise his written work, hoping to build confidence. It helped a little, but not enough to translate into verbal participation.

Then I introduced a strategy called "think-pair-share." Here's how it worked: I'd pose a question or problem. Students would spend a minute thinking about it individually. Then they'd turn to a partner and discuss their thinking. Only then would we share ideas with the whole class. For the first time, Luis spoke up. Not loudly. Not with confidence. But enough. He shared an idea with his partner, and then, when we moved to whole-class discussion, he was willing to say, "My partner and I thought..." This gave him a way to participate that felt safer.

Gradually, his confidence grew. He started raising his hand, just occasionally, but that was a start. By the end of the semester, he was contributing regularly to discussions. He wasn't the most talkative student in the class, and I didn't expect him to be. But he was present. He was participating. He was no longer invisible.

Creating Multiple Pathways for Participation

What I learned through Luis and many other students is that not every student expresses themselves the same way. Some are writers, some are talkers, some need time to process. Some are visual learners, some are kinesthetic, and some need to move around to think. Some are introverts who find large group discussions draining, while others are extroverts who thrive on them. Some have processing delays, some have anxiety, and some have learning disabilities that make traditional participation challenging.

For years, I was measuring participation primarily through verbal contribution in whole-class discussions. It was the most visible form of participation, but it was also the narrowest definition. By focusing only on that, I was silencing entire categories of learners.

I began offering multiple ways to participate: written reflections that students could share with the class. Collaborative projects where different students had different roles, some could be presenters, some could be researchers, some could be designers or builders. Digital submissions where students could record their thinking or respond to prompts in writing. Small group discussions where the stakes felt lower. One-on-one conversations with me where they could share ideas that they weren't comfortable sharing publicly.

Artistic or physical representations of ideas, creating something that shows their thinking without requiring verbal articulation.

I also started paying attention to when and how different students participated. Some students only spoke during small group work. Some only participated in writing. Some had brilliant ideas in one-on-one conversations that they'd never share publicly. I started to recognize all of these as legitimate forms of participation, and I started to intentionally create space for all of them.

The Culture of Respect and Risk-Taking

I also realized that I praised curiosity and creativity, but I wasn't always valuing the actual work of participation, which sometimes meant being wrong, uncertain, or confused. I started changing my feedback. Instead of just praising correct answers, I started praising questions, even questions that seemed tangential or confused. "That's an interesting question. Let's think about it." Instead of moving quickly past a wrong answer, I'd sometimes pause and say, "Let's explore this thinking. What made you think that?" I was trying to signal that the process of thinking, messy, uncertain, and sometimes wrong, was valued.

I made it a point to model mistakes. I'd solve a problem in front of the class, and I'd intentionally make an error, then catch it. "Oops, did you see that? I went wrong here. Let me rethink it." I'd share stories about times I'd been wrong or had to change my thinking. I'd admit when I didn't know something. I was trying to create a culture where mistakes were normal, expected, even valuable.

I celebrated unconventional thinking. When a student approached a problem differently than I'd taught, I'd stop and say, "I hadn't thought about it that way. Let's see if that works." When a student asked a question that challenged my thinking, instead of quickly defending my position, I'd say, "That's a great question. I'm going to think about that more." I was trying to signal that the classroom was a space where different perspectives were not just tolerated but genuinely valued.

Classroom culture matters profoundly. When students know their ideas are respected, they're more likely to take risks. When they know that mistakes are part of learning, they're more willing to try. When they see their teacher being humble and willing to learn, they're more comfortable with their own vulnerability. It takes consistent work to build this kind of culture, but the payoff is enormous.

The Power of Relationships

Relationships matter, too. The more I learned about my students' interests, their fears, their stories, their strengths, the more they trusted me, and each other. Even brief conversations before class built a connection. "I heard you have a soccer game after school. Good luck!" "I noticed you liked that book we read. I have another recommendation." "Is everything okay? You seem quiet today."

These small moments of connection signal to students that they're seen as individuals, not just as learners. When students feel known and valued, they're more willing to participate, more willing to take risks, more willing to be vulnerable.

I also started learning about students' lives outside of school. What music did they listen to? What were their family traditions? What challenges were they facing? What made them laugh? What were they passionate about? And I started bringing those things into the classroom. I'd use examples that connected to their interests. I'd celebrate their wins outside of school. I'd acknowledge their struggles with empathy. They started to feel like they belonged in my classroom, not just as students, but as whole people.

I also made sure to connect students. I created regular small group work, and I varied the groups intentionally so that different students worked together. I created projects where different students' strengths were needed. I facilitated conversations about what each student brought to the classroom. I started to build a community where students knew and valued each other, not just me as the teacher.

Addressing Participation Anxiety

Some students have genuine anxiety about participation, and simply creating a culture of respect isn't enough to overcome it. These students need additional strategies. I started working with anxious students individually. We'd talk about what specifically triggered their anxiety. Sometimes it was fear of judgment. Sometimes it was a worry about having the "right" answer. Sometimes it was just the vulnerability of speaking publicly.

We'd practice together. I'd ask them questions in one-on-one settings, giving them low-stakes opportunities to practice expressing their thinking. We'd work through some breathing techniques or grounding exercises to manage the physical symptoms of anxiety. We'd reframe participation as communication, not performance. The goal wasn't to sound smart or perfect; it was to contribute to the thinking of the group.

I also started using think-pair-share and other low-stakes strategies more regularly. These strategies reduced the pressure significantly. When you know you're going to talk to a partner first, the stakes feel much lower. When you know you can pass on a direct question, it feels safer. When you know you can submit written reflections instead of speaking, it feels more manageable.

For some students, I also used randomization in a way that felt fair. If I were cold-calling on students, I would make sure it was truly random and that I wasn't putting certain students on the spot more than others. Or I'd let students know ahead of time that they might be called on, so they could prepare. I was trying to reduce the anxiety of unpredictability.

Including the Quiet Voices

One thing I learned is that the quietest students often have the most interesting ideas. They're just not comfortable sharing them in large group settings. I began actively soliciting their input in quieter ways. I'd ask them to submit written reflections. I'd have one-on-one conversations where I explicitly asked for their perspective. I'd occasionally ask if they'd be willing to share a written response with the class (giving them the option to say no).

I also noticed that quiet students sometimes had brilliant ideas during small-group work, but those ideas never made it into the larger conversation because the louder voices dominated. So, I started having groups present not just their own conclusions, but also interesting ideas that came up during their discussion, even if they didn't end up being the group's official answer. This created a system where all ideas, including those from quieter students, could be heard.

I also created a "thinking notebook" where students could jot down questions, ideas, or observations they weren't comfortable sharing aloud. I'd collect these periodically and read through them. Some of the insights were remarkable. And I'd occasionally bring them back to the class: "I was reading through your thinking notebooks, and someone had a really interesting idea. Can I share it?" This gave voice to quiet thinkers without exposing them if they didn't want to be exposed.

Redefining Participation

Raising every hand isn't the goal. Including every voice is. Participation looks different for every student, and that's okay. The real challenge is making sure everyone feels they belong, that their contribution matters.

I started to reframe what I meant by "participation." Yes, verbal contribution in whole-class discussion was one form of it. But so was asking a thoughtful question in an email. So was collaborating effectively in a group project. So was contributing to an online discussion board. So was creating a piece of work that showed deep thinking. So was helping a peer understand a concept. So was showing up and engaging with the content, even if you weren't speaking out loud.

I started to assess participation more holistically. Instead of just tallying hand raises, I paid attention to engagement, effort, growth, and contribution in multiple forms. I noticed who was listening carefully and building on others' ideas. I noticed who was asking meaningful questions. I noticed who was helping peers. I noticed who was taking intellectual risks. I noticed who was showing up with genuine curiosity. These were all forms of participation, and they all mattered.

I also began giving students agency over how they demonstrated their understanding and engagement. Instead of requiring everyone to participate verbally, I'd offer options: "You can share your thinking out loud, or you can write it down, and I can read it. You can present to the whole class, or to a small group, or to me one-on-one. You can create something that represents your thinking. You can teach a peer. How would you like to show what you know?"

When students had a choice, and when the expectations were clear, more of them participated. And more importantly, they participated authentically in ways that matched their strengths and personalities. The introvert didn't have to become an extrovert to succeed in my class. The quiet thinker didn't have to compete with louder voices. The student with social anxiety didn't have to overcome their anxiety just to show me what they'd learned.

Reflective Prompts for Educators:

- How can I create a classroom where every student feels their voice matters, even if they're not the loudest in the room?
- What can I do to ensure that my classroom offers multiple ways for students to participate, catering to different learning styles and personalities?
- How can I actively encourage students like Luis, who are hesitant to speak, to find safe ways to express their ideas?
- What strategies can I use to celebrate mistakes and questions in my classroom, making students feel that it's okay to be uncertain?
- How can I give my quieter students the space and opportunity to contribute without forcing them into large group discussions?
- What can I do to build a classroom culture where all forms of participation—whether verbal, written, or creative are valued equally?
- How can I integrate more opportunities for low-stakes participation, like think-pair-share, to help anxious students build confidence?
- What steps can I take to ensure that even the quietest voices are heard and respected, without them having to compete with louder peers?
- How can I create an environment where students are empowered to choose how they demonstrate their learning, whether through speaking, writing, or other creative forms?
- What can I do to ensure my classroom supports students' unique strengths, allowing them to participate authentically in a way that matches their personality?

Conclusion: Listening Beyond the Lesson.

If there's one lesson I hope you take from this book, it's that teaching and learning start with listening. Not just to the answers, but to the pauses, the hesitations, the stories behind every raised or lowered hand. Not just to what students say, but to what they don't say. Not just to their words, but to their body language, their drawings, their silence, their choices about when to speak and when to stay quiet.

Listening is the most underrated skill in education. We talk about active listening as if it were a technique we could learn in a workshop. But real listening requires something deeper. It requires attention. It requires patience. It requires suspending judgment. It requires being genuinely curious about someone else's thinking, even when it's different from yours. It requires humility—the recognition that you might not understand, that you might be wrong, that the other person has wisdom you don't have.

When I listen to my students, I really listen. I hear so much more than the content of what they're saying. I hear their confidence or their fear. I hear whether they believe in themselves. I hear what they care about. I hear their creativity, their confusion, their insight. I hear the stories they're not telling in words, but they communicate through their choices, their engagement, and their struggle.

Every classroom is full of hidden stories: kids who are surviving more than we know, teachers who are giving more than they show, families who are supporting in quiet, invisible ways. The student who comes to school without breakfast is trying to focus on math while hungry. The student whose parents are pressuring them to be perfect. The student whose learning difference is undiagnosed and who thinks they're just stupid. The student who's grieving a loss. The student who's dealing with abuse. The student whose gender identity doesn't match what everyone expects. The student whose family is undocumented and living in fear. The student whose mental health is deteriorating. The student who's brilliant but socially isolated.

And parallel to these student stories are teacher stories. The teacher who stayed up until midnight grading and is running on two hours of sleep. The teacher who spent their own money on classroom supplies because the budget didn't cover it. The teacher who's dealing with their own health crisis while showing up to support their students. The teacher who feels like they're failing despite doing everything right. The teacher who's burned out but can't afford to leave. The teacher who loves this work but is barely surviving it.

Grades and test scores capture so little of what's actually happening in a classroom. A report card might say that a student got a B in math, but it doesn't say that they learned perseverance by struggling with a difficult concept. It doesn't say that they asked for help for the first time. It doesn't say that they realized they could do hard things. It doesn't say that they found a community where they felt accepted. It doesn't say that a teacher saw potential in them that they didn't see in themselves.

The real work, the real transformation, happens in the spaces between the moments when someone takes the time to notice, to care, to listen. It happens in the hallway conversation where a student finally admits they're struggling. It happens in the office during lunch, where a teacher breaks down, and a colleague just sits with them. It happens during a small-group discussion when a student hears their peers respect their ideas. It happens during a one-on-one conference, when a parent learns that their child is more than their grades. It happens in the moment when someone realizes they're not alone in their struggle, that other people understand, that there's hope.

Who Are the Best Teachers?

The best teachers aren't the ones with the flashiest lessons or the neatest classrooms. They're not the ones who have it all figured out or who never struggle. They're the ones who pay attention. They're the ones who see the students who think they're invisible. They're the ones who believe in the possibility of growth even when the evidence is thin. They're the ones who remember that every student is more than a number, more than a row in a seating chart, more than a test score.

The best teachers are the ones who listen. Who ask real questions and wait for real answers. Who admit when they're wrong. Those who learn from their students. Those who show vulnerability alongside strength. Who create space for all kinds of learners and all kinds of thinking. Who understands that teaching is fundamentally relational work, that it's not about what you deliver, but about who you become together with your students.

The best teachers are also the ones who take care of themselves. Those who have boundaries. Who asks for help? Who know their limits and communicate them clearly. Those who celebrate small wins. Who builds community with their colleagues? Who remember that they can't pour from an empty cup, and that taking care of their own wellbeing is taking care of their students.

The Role of Families and Communities

Parents, caregivers, and communities, your role matters, too. Ask questions, listen more than you talk, celebrate effort as much as achievement. When your child comes home and says they don't understand something, instead of immediately stepping in to fix it, ask questions: "What did you try? What made sense to you? What were you confused about?" This teaches problem-solving and persistence.

Partner with teachers, not just to raise grades, but to raise spirits. A kind email to a teacher can sustain them through a difficult week. Recognizing their effort and sacrifice matters. And when there's a problem, approach it as a partner, not an adversary. "My child said something happened in your class. I'd like to understand what's going on." This is very different from "You did something wrong and need to fix it."

When we work together, families, teachers, students, and communities, we give every student a better chance to be seen, to be heard, to be known. We create a culture where belonging is the default instead of something that has to be earned. We help young people understand that they matter, that their voice matters, that their presence in the world makes a difference.

Teaching Has Changed Me

Teaching has changed me profoundly. It's made me more patient, more empathetic, more aware of my own limitations and strengths. It's taught me to see people—really see them—instead of seeing roles or categories. It's taught me that there's almost always more complexity than what meets the eye. It's taught me that change is possible, even when it seems impossible. It's taught me that what we do matters, that the relationships we build matter, and that the time we take to really listen to someone matters.

The stories my students have shared, sometimes in words, sometimes in gestures, have shaped the way I see the world. They've taught me about resilience and kindness and creativity. They've shown me what it looks like to keep trying even when you're scared. They've demonstrated the power of believing in someone. They've reminded me that every person has stories worth hearing, wisdom worth considering, experiences worth honoring.

I've learned that the real curriculum is human. The real lesson is connection. The real measure of success isn't a test score; it's whether every student in a classroom feels like they belong, like their thinking matters, like they have the capacity to learn and grow and change the world.

A Call to Action

So, here's what I'm asking you to do, starting today: listen harder. Look for the stories that hide behind silence. Notice the student who doesn't raise their hand but whose eyes light up during small group work. Notice the teacher who stays after school to grade papers. Notice the parent who's doing everything they can with the resources they have. Remember that every raised hand, every quiet voice, every small risk is a chance to change a life, including your own.

Create space in your classroom for all ways of being and thinking. Don't just tolerate difference; celebrate it. Don't just include students; make them feel genuinely welcome. Don't just answer questions; ask real questions and listen to the answers. Don't just teach content; build relationships. Don't just give grades; tell stories about growth and effort and becoming.

If you're a teacher, take care of yourself. Set boundaries. Ask for help. Build community with colleagues. Remember that you can't pour from an empty cup. And remember that your work—the real work of seeing students, of believing in them, of showing up for them day after day—that matters more than any test score or evaluation.

If you're a parent, ask questions and listen. Partner with teachers. Celebrate effort and growth. Tell your child that you believe in them, that mistakes are part of learning, and that their worth isn't tied to their grades. Help them see that the real measure of success is whether they're learning, growing, becoming more of who they're meant to be.

If you're a student reading this, you are enough. Your voice matters. Your thinking matters. Your presence in the world matters. The struggle you're experiencing isn't a sign that you're failing; it's a sign that you're learning. The mistakes you make are how you grow. The teachers who believe in you, the families who support you, the peers who accept you—hold onto those. They're the real curriculum. They're what matters.

Reflective Prompts for Educators, Parents, and Communities:

- How can I listen more deeply to the hidden stories around me?
- How can I listen beyond the words, tuning into the unspoken stories and silences of my students?
- What can I do to ensure every student feels their voice matters, even if they're not the first to speak up?
- How can I celebrate all forms of participation, whether verbal, written, or creative, in my classroom?
- What actions can I take to build deeper connections with my students, making them feel seen and heard beyond grades?
- How can I focus on relationships and growth, not just content, to help students feel valued and capable?
- What can I do to foster a classroom culture where mistakes are seen as learning opportunities and risks are encouraged?
- How can I actively listen to my students' challenges and turn those moments into chances for growth?
- What steps can I take to model vulnerability, showing students it's okay to not have all the answers?
- How can I celebrate effort and progress, not just the outcome or grade?
- What can I do to prioritize my well-being as a teacher, setting boundaries and building support so I can give my best every day?

Final Thoughts

Teaching is far from a simple profession; it's a dynamic, complex, and sometimes overwhelming adventure. It's messy, challenging, and, yes, occasionally heartbreaking. But above all, it's transformative. It's where humans connect in the most profound ways, bridging differences, sharing stories, and learning not just knowledge, but how to truly *be* in the world. It's where teachers uncover their own capacity for compassion, resilience, and unwavering commitment to those they serve.

The stories shared throughout this book are real. These are not fairy tales with tidy conclusions or guarantees of success. Instead, they reveal the messy, real-life challenges faced by students and teachers alike. Yet within these challenges lies something extraordinary: the courage to see students for who they truly are, to believe in their untapped potential, and to carve out space for them to grow and flourish, despite the odds.

As you move forward, whether as a teacher, parent, or community member, carry these powerful stories with you. Let them ignite something within you.

Behind every behavior, there's a story. Behind every struggle, a person is doing their absolute best with what they've been given. Behind every question, there's an opportunity to connect, to inspire, and to change a life forever.

Listen. See. Believe. Support.

These are the verbs that fuel education. They are the heart of what transforms classrooms, uplifts lives, and ignites passions. These aren't just actions, they're powers. They turn challenges into triumphs and make an impact far beyond what we can measure with grades or assessments.

So, as you step forward into this relentless, beautiful work, remember this:

The real work is just beginning. The impact we have as educators, mentors, and supporters isn't defined by perfection; it's defined by presence, by the unshakeable belief that every student, no matter how quiet, no matter how invisible, has the power to rise, to thrive, and to transform the world.

So, keep showing up with energy, with passion, and with fire. Keep listening with an open heart, seeing with a clear vision, and believing without hesitation in the incredible potential of every single student. The world is waiting for them and for you to help them discover what's possible.

Made in the USA
Las Vegas, NV
19 September 2026

cf2ec785-fcf2-4d16-82e5-27c6e563b106R02